THE
HUMBLE
THEORY OF
EVERYTHING

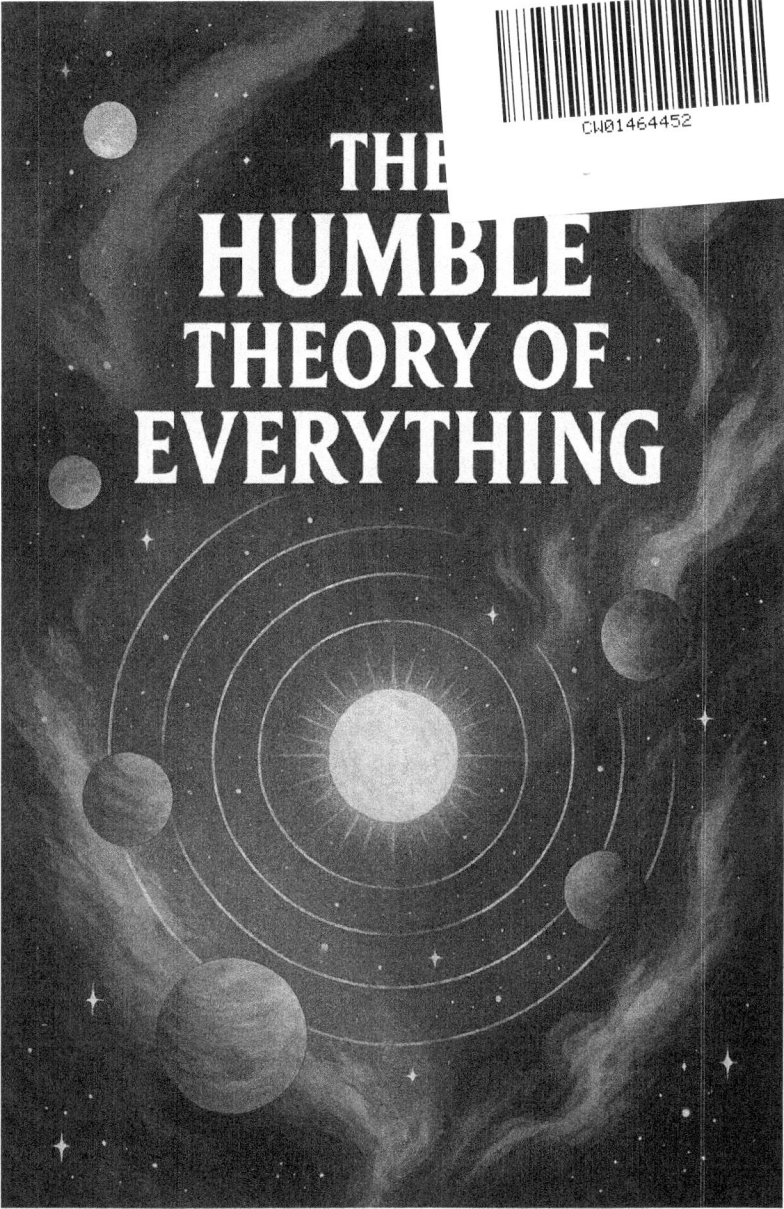

Index:

The Big Bang and Existence Within a Black Hole.

I want to begin at the only true beginning. Not the birth of Earth. Not the dawn of stars. Not even the expansion of the universe as we know it. I want to push further back — before galaxies, before atoms, before light itself. I am speaking of a void so complete that even the word "nothing" feels too heavy to describe it. A state without time, without matter, without energy, without motion. No space to measure, no clock to tick, no memory to hold it. A silence more absolute than death.

And then, suddenly, from this void came the first soundless thunder. The Big Bang. An explosion without a place to explode. A burst of energy without any background to emerge against. In a fraction of a fraction of a second, all that we call existence was unleashed — matter, light, the fundamental forces, the seeds of galaxies, and eventually, the possibility of life. That one moment, the singular instant of creation, remains the most profound mystery in all of science.

We have measured its afterglow — the faint microwave radiation still humming across the sky, the so-called fossil light of creation. We have calculated its rapid expansion, modeled its inflation, and traced the story of atoms coalescing into stars. And yet, beneath all this knowledge, the most basic question has never been answered: where did the Bang come from?

Most accounts stop at the bang itself. Some scientists suggest it was a quantum fluctuation, a random ripple in a sea of uncertainty. Others invoke multiverse theories, imagining that our Big Bang was only one bubble among many, frothing eternally. Religious traditions see it as the hand of God, a fiat of divine will. But none of these answers feel entirely satisfying to me. I believe there is another possibility, one that ties together what we see in the skies with what we already know about gravity, matter, and collapse.

What if the Big Bang did not come out of nothing? What if it came out of something we already know exists, something we have already observed? What if the Big Bang came from inside a black hole? This is the question that sparks my hypothesis.

To most, it might sound absurd. Black holes, after all, are cosmic sinks, devourers of matter and light. They swallow everything that falls into them and give nothing back. They are thought to be endpoints — the dead ends of stars, the final punctuation marks of galaxies. How could something that collapses everything into silence also give birth to everything we know?

And yet, the more I sit with this question, the more sense it makes. We know black holes are real. We have watched stars orbit invisible centers with impossible speed. We have measured the X-ray jets from accretion disks spiraling into nothingness. We have even photographed their shadows — a ring of fire around an absence. If black holes exist, and if the Big Bang happened, then perhaps the two are not separate mysteries but one.

I imagine it like this: space itself is not flat and empty but flexible, textured, and dynamic. The old classroom analogy shows us a rubber sheet stretching under the weight of mass, but I see something more intricate. A sheet with dips, ridges, folds, and hidden terrain — a cosmic fabric sculpted not by matter, but by gravity itself. In this vision, gravity comes first. Gravity is the architecture, and matter is the filling. This would explain the mystery of dark matter: invisible wells of gravity pulling on galaxies, wells we can measure but not yet see filled. The gravity is there. The matter is only its shadow.

If this is true, then the Big Bang may have been less of a singular miracle and more of a recurring process — the natural consequence of a gravitational funnel collapsing so completely that it tore through its own sheet. A black hole, but one pushed to an ultimate threshold. When that tear occurred, what had been compressed beyond imagination burst into an entirely new space, a fresh canvas, a dimension without history. To those who inhabit it — to us — it would feel like creation itself. To the parent universe, it would look like the silent vanishing of matter into a black hole.

This is the twist I want to explore. The Bang as the other side of collapse. The beginning of everything as the inside-out of a black hole.

The Gravity-First Universe.

If I want to understand the beginning, I need a framework sturdy enough to hold the weight of the cosmos. For me, that framework begins with gravity — not as a byproduct of matter, but as the primal force that shapes matter itself. Most of us grew up with the schoolroom demonstration: a rubber sheet stretched taut, a heavy ball placed on it, and marbles rolling toward the depression. This was Einstein's gift — the picture of spacetime bending under mass. Matter tells space how to curve; space tells matter how to move. It is elegant, simple, and it works. But I think there is a missing step.

I don't see matter as the cause. I see gravity as primary. The sheet itself is not neutral until mass arrives. It already has texture. It already has dips, ridges, folds, and wells. Matter doesn't make them — it merely falls into them. Gravity comes first. That single shift in perspective changes everything. If gravity wells already exist, then dark matter becomes far less mysterious. Right now, scientists see galaxies rotating as though they are being pulled by invisible hands. They cannot account for the motion with the matter we can see, so they propose there must be hidden mass — dark matter — spread invisibly through the cosmos. Billions have been spent searching for these missing particles, and yet not one has been captured.

But if gravity precedes matter, the search may be misdirected. We are not missing particles. We are seeing wells that have not yet been filled. Invisible pits in the sheet of spacetime, waiting for matter to tumble into them. Gravity wells pull at galaxies long before they hold any substance of their own. The gravity is real, measurable, undeniable. The matter is optional. Think of it like a landscape before a flood. Valleys and ravines exist whether or not they are filled with water. When the rains come, rivers and lakes form. But even in drought, the land's shape is there. I believe the cosmos is the same. Gravity carved the terrain first. Matter is the rainfall that pools in its low places.

Bowls, Cups, and Cones: The Shapes of Gravity.

If space is not flat but textured, then it makes sense to picture its terrain as varied — like hills, valleys, basins, and cliffs. Gravity, in this framework, is the sculptor. Matter does not shape space; instead, matter rolls into the spaces already carved. And just as water fills valleys according to their depth and shape, matter fills gravitational wells according to their geometry.
I think of these wells in three main categories: bowls, cups, and cones.

Bowls are the wide depressions, shallow but broad. They do not concentrate matter in a single point, but spread it widely across their area. These bowls are the cradles of galaxies. Stars scatter within them the way pebbles scatter across the bottom of a large basin. They spin in shared orbits, each star responding not to its own gravity alone but to the shape of the larger bowl. This explains why galaxies form spirals and clusters — not because they invented their own order, but because they are following the lines of the bowl they inhabit. In this sense, galaxies are not self-contained objects but the visible tracings of an invisible depression. Their shape and rotation are signatures of the bowl's contour. Dark matter halos, in my model, are not halos of mysterious particles at all, but the unlit edges of these vast bowls. The matter is simply mapping the terrain.

Cups are smaller, deeper, and more focused than bowls. Where bowls disperse, cups concentrate. These are the seedbeds of stars. Hydrogen gas falls into them, compresses under its own weight, and ignites into nuclear fusion. Stars, quasars, and compact systems sit in cups like molten beads at the bottom of a ladle. Cups explain why stars form in clusters or within certain bands of galaxies. They are not sprinkled randomly but born in regions where cups dot the terrain. Quasars — those hyper-bright beacons in the distant universe — are cups so intense that they trap entire galaxies' worth of fuel, funneling it into luminous jets.

And then there are the cones. Rare, narrow, and infinitely patient, cones are the ultimate gravitational shape. Unlike bowls or cups, which balance matter in broad or moderate depressions, cones are funnels with no end point — channels that draw everything downward toward a single focus. Matter that enters a cone does not spread or ignite. It spirals. Slowly at first, then faster, and faster still, until the fall becomes unstoppable. Eons pass as the funnel feeds, but it never surrenders its pull. Eventually, the matter condenses into a point so dense that space itself cannot contain it. The cone collapses into its own tip. This is the birth of a black hole.

The Clown Horn at the End of the Universe. The metaphor may sound ridiculous, but sometimes the silly image captures the truth better than solemn words. Imagine an old clown horn — a long, narrow cone ending in a tiny rubber bulb. Now replace the bulb with a singularity, a sphere of infinite density. That is the structure of a cone-shaped well when it collapses.

"Meep meep!" might be funny, but it hides a profound fact: the horn is the funnel, the tip is the singularity, and once matter reaches that point, it vanishes from our visible sheet. From the outside, all we see is the horn glowing faintly around its edges — an accretion disk — while the bulb swallows everything with silence.

But unlike a clown's toy, the singularity does not stay still. Its gravity grows with every bite. It expands its influence, pulling in more matter, dragging the very fabric of space with it. The funnel deepens, the pull accelerates, and eventually the singularity becomes so voracious it threatens to tear the sheet itself. This is not just collapse. This is potential creation.

The Twist: Where Does It All Go?

The beauty of the universe is that it rarely leaves loose ends. Every question we ask eventually folds back into the fabric itself, if we look closely enough. And so I must ask the most obvious question about black holes: where does it all go? We see the funnels. We watch stars circle their edges at a fraction of light-speed. We measure the X-ray jets as matter heats and tears apart on the way down. We even have the eerie photographs — dark voids ringed with fire, proof that the abyss is not just theory. But we never see what happens beyond the horizon. From our side, it is silent.

Traditional physics offers a bleak answer: nothing goes anywhere. Matter collapses into a singularity of infinite density, all information lost, time and space pinched out of existence. End of story. But I find that answer unsatisfying, not because it is frightening, but because nature rarely builds dead ends. Everywhere we look, the cosmos recycles. Stars die, but their matter becomes the stuff of planets. Forests burn, but their ash fertilizes the soil. Even our cells renew, shedding and replacing, one cycle feeding the next. Why should black holes — the most extreme engines of nature — be exceptions?

Some have proposed "white holes," the theoretical opposites of black holes. Instead of swallowing, they spit. Instead of silence, they pour. Matter and energy expelled in jets of creation, perhaps even forming new universes. But there's a problem: we've never seen one. Not a hint, not a shadow. And the more we understand about black holes, the less plausible their opposite seems. But opposites are not always needed. Sunlight does not require moonlight. Thunder does not demand silence. The universe does not always balance itself by creating a mirror image. Sometimes it balances by transformation. So here is my hypothesis: black holes don't end in singular silence. They end in rupture.

When enough mass and energy are crushed into the funnel, when the cone has gathered eons of collapse into its narrow throat, the structure can no longer contain itself. Space and time themselves are pulled beyond their limits. A tear forms. A rip in the sheet. A dimensional puncture. And through that rip, all the stored matter and energy burst into somewhere else. Not into our universe. Into a new one. This is the birth of a Big Bang.

From the outside, we see only the collapse. Stars vanish, light dies, and the horizon deepens into shadow. But from the inside of the tear, the perspective is reversed. There, it is not collapse but expansion. The singularity is not a prison, but a womb. What looks like the death of matter here is its first breath there. The rip is both grave and cradle, both end and beginning.

This explains why black holes sometimes brighten and dim. Before the tear, they are swollen with matter, dense enough that even light struggles to escape. After the tear, their mass has offloaded into the new dimension. The singularity grows lighter, its grip loosens, and suddenly more light slips around the edges. It is as though the black hole exhales after eons of holding its breath. And then, inevitably, the funnel begins gathering again. In this model, the universe is not a one-time miracle but a recursive process. Big Bangs are not isolated events, but natural consequences of gravitational funnels reaching their limit. Creation is not a singular past but an ongoing cycle.

And if this is true, then every black hole is a potential cosmos in waiting. Every funnel, no matter how small it seems now, is building toward a threshold. Somewhere, beyond our sight, the collapse may already have triggered. Somewhere, a new sheet of spacetime may be blooming. And in that new sheet, perhaps beings are already looking up, asking the same questions we ask now: where did their beginning come from? The twist, then, is not that black holes are destroyers. The twist is that they are doorways. Not to white holes, not to mirrored opposites, but to birth itself. Each one is a Big Bang seed, waiting for its moment.

Cycles and Timing: One Bang Per Customer.

If black holes can tear new universes into being, then why don't we see Big Bangs happening all the time? Why just the one behind us? Why not fireworks bursting across the sky of the multiverse every other Tuesday? The answer, I believe, lies in scale — in the patience of the cosmos itself. We measure time in minutes, years, lifetimes. The universe measures in epochs, in spans so long that language barely captures them. A star may burn for billions of years before dying. A galaxy may swirl for tens of billions before reshaping. And a black hole, that slow and patient funnel, takes eons upon eons to gather enough mass to rip through its own fabric.

Imagine a mountain-sized hourglass, its grains of sand made of stars. Each grain takes centuries to fall, but they do fall, one by one, pulled into the funnel. Galaxies feed their cores, quasars burn themselves out, wandering matter drifts inward, and over inconceivable time, the black hole swells. Then, at last, when the threshold is reached, the singularity punctures through. In that moment — and only in that moment — a new universe is born.

It is not that multiple Big Bangs don't happen. It is that they happen so rarely, on such an immense timescale, that within any single sheet of spacetime, only one is ever visible. By the time a universe like ours comes into being, its siblings are either far beyond our horizon or yet unborn, their funnels still gathering their fuel. So, one bang per dimension. One bang per lifetime of a universe.

The joke almost writes itself: "one bang per customer." A playful phrase, but it carries truth. Not because the cosmos is stingy, but because the mechanics of gravity demand patience. Black holes are engines of inevitability, but inevitability takes time. And in that patience, something beautiful unfolds: the cosmic lineage. Each sheet inherits its existence from the black hole of a parent sheet. Each parent waits eons to gestate its offspring. Each offspring will in turn develop its own funnels, its own tears, its own Big Bang. A recursive family tree drawn not in branches but in tunnels, spiraling across the multiverse like galaxies of galaxies.

This explains why we perceive only our Big Bang. It is not unique — it is simply the one that birthed our sheet. Other bangs are happening, or will happen, but they are sealed behind horizons we cannot cross. They are the distant lights of other tunnels, too far away, too slow to emerge in the timescales of our lives.

The irony is that we live in the middle of infinity but only ever see one beginning. It is the illusion of centrality — the same mistake that once led humanity to think Earth was the center of the cosmos. We assumed our Big Bang was the Big Bang, the singular event. But in truth, it may only be one in an endless series, one branch of a recursive lineage of universes. Patience is the rule of creation. Not fireworks every day, but a single thunderclap per dimension. And that thunderclap, stretched across billions of years of unfolding, is what we call our universe.

Multiversal Structure: Sheets in the Void.

If black holes are wombs of new universes, then where do those universes go? Do they stack like pages in a book, or do they scatter like seeds in a wind? I believe the answer is stranger and more elegant: each universe is a sheet, floating in a greater void — a multiversal ocean where dimensional layers drift, spiral, and seed new worlds.

The sheet analogy still works, but now it must be scaled up. Imagine our universe not as the totality of existence but as one great, flexible surface stretched across the void. This surface is alive with curves and folds, each fold capable of birthing black holes. Every time one of those wells grows heavy enough to rip, it punctures through the sheet and births another sheet below it. A new universe unfurls, tethered to its parent by the funnel of the black hole.

In this picture, universes are not isolated bubbles. They are linked in chains, each child trailing from a black hole in its parent, and each parent spiraling in motion through the greater void. The multiverse is not a neat stack of pancakes. It is a school of fish, a galaxy of galaxies, a recursive lineage moving in corkscrews across something bigger than space itself.

To visualize it, return to the solar system. We often imagine planets orbiting the sun in flat ellipses, but in reality, the whole system is moving forward through the galaxy. The sun is a locomotive, and the planets spiral behind it like ribbons in a wake. The orbits are corkscrews, not circles. I believe our universe — our entire cosmic sheet — does the same. It drifts forward through the greater void, and trailing behind it, birthed from its black holes, are smaller sheets: its children. They spiral in its wake, each carrying its own physics, its own histories, its own Big Bang.

This spiral model explains both motion and connection. It explains why universes may never collide: they follow the gravity of a larger anchor, like planets tethered to a star. And it explains why every universe feels central to itself: each one was born from its own tear, its own tunnel, and carries on its own trajectory. We are children of one black hole, just as surely as stars are children of supernovae.

The beauty of this recursive structure is its lineage. Just as DNA carries forward a genetic code, black holes carry forward the energy and matter of a parent universe into the child. Each universe is not a random burst but an inheritance. Patterns may repeat — constants, ratios, even laws of physics may echo — because they are encoded in the funnel that birthed them. And just as children differ from parents, each new universe may contain slight variations: new configurations of matter, new strengths of gravity, new landscapes of possibility.

So what we call the multiverse may not be a chaotic infinity. It may be a family tree. A living genealogy of dimensions, spiraling endlessly through the greater void, each one feeding the next. And in that cosmic spiral, our own universe is but one branch — a tunnel-child of something larger, a parent to others yet unborn. As above, so below.

The Edges of Our Universe: The Dark Tunnel Horizon.

When we ask about the edges of the universe, we tend to imagine a boundary — a wall of stars, a rim of galaxies, perhaps even a line beyond which there is nothing. But that picture is too flat, too final. The edges of our cosmos, I believe, behave more like an event horizon. Not the event horizon of a single black hole, but of the universe itself — the collective boundary of the sheet we inhabit.

Think of a black hole's horizon: the point beyond which no light returns. To an outside observer, it appears as a shadowed edge, a glowing ring where spacetime curves back on itself. What if our universe has something similar, not for a collapsed star but for the cosmos as a whole? When we look outward far enough, across billions of light years, what we see may not only be the cosmic microwave background — the "fossil glow" of the Big Bang — but the light skimming around the curve of the tunnel that birthed us. I call this the dark tunnel horizon.

This reframing changes everything. In standard cosmology, the background radiation is leftover heat, diffusing evenly across space. But in the black-hole-birth model, it becomes something more: the curved rim of the dimensional funnel that launched our reality. Its uniformity makes sense not as randomness smoothed by cosmic inflation, but as geometry — the symmetry of a tunnel whose walls look the same in every direction.

If this is true, then our universe is finite, but not with hard edges or borders. It is bounded by curvature. We live inside a tunnel, not an endless plain. The faint glow at the farthest reaches is not the afterburn of creation, but the shimmer of the structure itself — the echo of the rip that opened this sheet.

And here's where recursion reveals its elegance. Every black hole that grows massive enough tears its sheet and births a new tunnel.

Every tunnel is a universe. Our cosmos is not a lone expanse, but one pipe in a vast, spiraling network of pipes — each born from the collapse of the last, each moving through the void like corkscrews behind a sun. A lineage of tunnels, endlessly recursive, yet always finite in their own frame.

From the inside, we cannot leave. We never escape the tunnels. They are the architecture of reality itself. But by looking outward, by studying the faint horizon at the edge of vision, we glimpse the shape of our container. And in that glimpse, we learn: the end is not a wall, but a curve. The cosmos itself has an event horizon.

That boundary we see, the strange glow mapped by satellites and studied in detail, may be more than ancient heat. It may be the dark tunnel horizon — the reminder that we live inside a recursive structure of unimaginable scale, and that our story is one tunnel among many.

Restating the Premise.

Let us begin with a reversal of the assumption most of modern physics takes for granted. In the standard model of the universe, matter is the sculptor, and gravity is the sculpture. Mass, we are told, bends spacetime around it, creating the curvature that we call gravity. This has been the dominant framework since Einstein's general relativity replaced Newton's invisible "force of attraction" with the geometry of spacetime itself. But here, I propose something even more fundamental: that gravity is not sculpted by matter at all. Gravity is the terrain itself, the landscape of spacetime that was present before matter arrived. Matter does not create the hills and valleys—it merely falls into them.

This subtle shift of perspective changes everything. Instead of viewing matter as the cause of curvature, we now imagine curvature as the precondition for matter. Gravity is not a secondary effect; it is primary. It is the stage upon which the atomic actors perform, the floor upon which the play of physics unfolds. Mass is not the creator of gravity; mass is the response to gravity's invitation.

In this model, gravity wells are not afterthoughts of clustered matter but ancient contours in the cosmic sheet. Picture spacetime as a landscape of ridges, bowls, and tunnels, etched into existence from the very beginning. Matter, when it emerges, does not carve these wells but rather pools within them—like water flowing into a basin. Some wells are shallow, spreading mass across broad galactic scales. Others are narrow, concentrating stars or quasars. And some, the rarest and deepest, are cone-shaped funnels where all matter is pulled relentlessly toward a single point, eventually forming black holes. But the crucial point is this: the wells exist before the water. The terrain predates the traveler.

If this premise is true, then our understanding of the nuclear forces must be revisited. In the traditional hierarchy, gravity is considered negligible in the subatomic realm. The strong nuclear force dominates at the scale of quarks, binding them into protons and neutrons. The weak nuclear force governs particle decay and identity shifts. Gravity, by comparison, is so faint at these scales that physicists often treat it as irrelevant—an annoyance to be ignored in particle physics calculations. But this dismissal depends on assuming gravity is caused by mass. If, instead, gravity is the primordial terrain and matters the secondary arrival, then the nuclear forces do not exist in isolation. They are not rulers imposing their will on a neutral stage. They are responses—behaviors arising within the terrain gravity has already set.

This reframing casts the strong and weak forces in a new light. They are no longer independent artisans shaping the atom, but craftsmen working with tools gravity has already provided. Their behavior is subtly but inexorably influenced by the curvature of the wells they inhabit. In steep wells, the strong force may bind more tightly. In warped geometries, the weak force may tunnel differently, probing seams in spacetime itself. The constants we treat as universal may in fact be local, their values tuned not by abstract law but by the hidden terrain of gravity that lies beneath.

It is a shift from a universe where forces rule to one where forces respond. Instead of imagining matter and its forces as sovereign, dictating how spacetime must bend, we imagine spacetime already shaped—gravity already etched—long before a single proton ever condensed. In this view, physics is resonance. The nuclear forces play the melody, but gravity has already written the key signature.

Strong Force and Gravity Wells.

The strong nuclear force is the most powerful known interaction in the universe. It is the glue that holds quarks together inside protons and neutrons, and in turn binds those protons and neutrons into nuclei. Without it, no atoms could exist; without atoms, no chemistry; without chemistry, no life. In standard physics, the strong force is treated as an isolated titan, dwarfing gravity so thoroughly at the subatomic scale that gravity is dismissed as irrelevant. But if gravity precedes matter—if it is the terrain into which all particles fall—then even the strong force must operate within the contours of this pre-existing landscape.

Imagine the birth of a proton not in some flat, neutral arena, but in the warped floor of a gravitational dip. The well is already there, curved and contoured, shaping the geometry of space itself. When quarks begin to bind, their gluon field lines do not stretch in a simple, straight lattice. Instead, they bend, arc, and condense according to the terrain beneath them. The "constants" of the strong force may not be truly constant, but locally modulated by the gravity architecture in which they operate.

This opens the door to a profound implication: the binding energy of nuclei may vary depending on the steepness of the local well. In shallow wells, the strong force behaves as we have measured it on Earth — binding with a certain predictable energy, producing stable isotopes up to iron, beyond which binding becomes energetically costly. But in deeper wells — such as those near neutron stars or the event horizons of black holes — the binding efficiency may be slightly boosted. Quarks could "snap" into more stable arrangements, exotic isotopes could endure longer, and elements more complex than those found naturally on Earth might form.

Consider the environment of a neutron star. Here, gravity is so intense that atoms collapse into a sea of neutrons, with protons and electrons forced into fusion by sheer compression. Standard physics explains this as pressure overwhelming the Pauli exclusion principle. But in the gravity-first framework, the explanation deepens: the local well reshapes the strong force itself, making confinement more favorable and shifting the balance of stability. Matter in these regions is not merely crushed — it is "gravity-forged," emerging in forms that may not even be possible elsewhere in the universe.

The same principle applies near black hole accretion zones. At the edges of the event horizon, particles whirl in extreme gravitational gradients. If gluon confinement curves differently in these warped terrains, then the nuclear chemistry of such regions could differ dramatically from our own. The periodic table we know may be only a local expression of element formation; elsewhere, under the influence of sharper wells, matter may arrange into new patterns, new isotopes, perhaps even entirely new classes of stable nuclei.

Another implication is anisotropy — direction-dependent binding. On Earth, the strong force appears isotropic, the same in every direction. But within a warped well, gluon field lines may not spread symmetrically. They may condense more tightly along the axis of curvature, producing particles with unusual directional properties or favoring certain binding configurations. If this is true, then the diversity of matter across the universe is not only a story of temperature and pressure but also of gravitational architecture. The periodic table is not universal — it is local, written into the terrain where it was forged.

We can think of this as the geology of spacetime at the quantum level. Just as mountains, rivers, and valleys shape the distribution of ecosystems on Earth, gravity wells shape the distribution of atomic possibilities. A shallow well produces familiar terrain: hydrogen, helium, carbon, iron. A deep well creates a harsher environment where only exotic "lifeforms" of matter can endure. And the deepest wells, the cones collapsing into singularities, may give rise to elements and binding patterns we cannot yet imagine — matter that belongs to the interiors of black holes, never to drift outward into our telescopes.

What follows is a subtle but radical idea: that the strong force is not truly independent, but resonant. It responds to the contours of the gravity that pre-existed it. The "constants" we measure in laboratories are not eternal laws but local properties of the terrain we happen to inhabit. If we could measure the strong force deep within a neutron star or just outside a black hole, the numbers might differ. Gravity does not merely bend the paths of planets or light beams — it shapes the very rules by which particles hold together.

This perspective transforms our understanding of cosmic matter. Galaxies, stars, and planets are not just organized by gravity at large scales; their very substance — the nuclei in their cores — may bear the imprint of the wells in which they were forged. Gravity is not just the architect of structure. It is the hidden artisan of matter itself.

Weak Force as the Tunnel Key.

The weak nuclear force is often introduced in physics classrooms as the quiet cousin of the strong interaction. It governs radioactive decay, neutrino interactions, and the subtle transformations of particles that flip identities in fleeting instants. Compared to the mighty strong force, which glues quarks together, or electromagnetism, which shapes the visible world, the weak force feels almost ghostlike — invisible, elusive, rarely noticed except in the delicate ticking of radioactive half-lives. But in a gravity-first universe, where spacetime itself is the terrain and matter is only the traveler, the weak force takes on an entirely new role: not as a weakling, but as a key. A tunnel key.

At its heart, the weak force is about transitions. It allows a neutron to turn into a proton, an electron, and an antineutrino. It lets quarks change flavor, shifting the fundamental identity of matter itself. Unlike the strong force, which is about stability and binding, the weak force is about change and passage. In the standard model, this is framed mathematically in terms of W and Z bosons mediating decay. But if gravity pre-shapes the terrain, then the weak force is the process that tunnels through that terrain. It does not merely govern random radioactive breakdowns. It probes the seams in spacetime itself.

In our laboratories, decay rates are considered constant. A half-life is treated as universal, untouched by the environment except for relativistic time dilation at high speeds or strong gravity fields. But if the weak force operates within pre-structured gravity wells, then decay is not only stretched by relativistic clocks but tuned by local curvature.

Inside a steep gravitational dip — such as the warped geometry near a neutron star or within a collapsing black hole's accretion zone — the probability landscape for tunneling shifts. Particles that would ordinarily decay in predictable intervals may linger longer or vanish faster, depending on the terrain. Decay constants may, in fact, be local properties, modulated by the topology of space itself. The weak force is not "weak" in these regions; it is resonant with the curvature, its behavior bent by the same invisible terrain that channels matter into wells.

Neutrinos — the most elusive of particles — provide a powerful illustration of this role. Billions pass through your body every second, barely interacting, slipping ghostlike through planets and stars. But in the gravity-first model, neutrinos are not truly indifferent wanderers. Their flight paths and interaction probabilities may be subtly biased by curvature.

In deep wells, neutrinos may be nudged into greater alignment with the funnel's structure, traveling not randomly but as if following an invisible compass needle pointing along the gravitational field lines. In this sense, neutrinos could be the universe's hidden "surveyors," tracing the terrain of spacetime itself. Perhaps this is why neutrino detection is so staggeringly difficult — because we measure them only in the flat shallows of Earth's gravitational well. Near the extreme dips of black holes, their role may be far more dramatic, carrying signals across the very seams of reality.

At the most extreme boundaries — the event horizon of black holes, or the points where spacetime itself fractures into new domains — the weak force may reveal its most profound function. If the strong force is about binding matter within a shaped terrain, the weak force is about allowing matter to slip between terrains. It is the pathfinder at the edge of dimensions.

Imagine the collapse of a star. As matter cascades toward singularity, particles undergo transformations at staggering rates. Quarks flip, leptons emerge, neutrinos stream out in floods. In these crucibles, the weak force is not just a decay mechanism — it becomes the negotiator between states of being. It may even serve as the mechanism that helps matter transition from one sheet of spacetime into another, a tunneling process magnified at cosmic scale. The weak force, in this view, is the agent that senses and exploits the seams gravity creates when it tears.

This frames the weak force not as an afterthought, but as the universe's subtle locksmith. It carries the keys that unlock transitions across scales: from one particle identity to another, from one nuclear state to another, and — in the extreme — from one dimension of spacetime to another.

When seen together, the strong and weak forces are not opposites but complements within the gravity-first framework. The strong force responds to terrain by binding matter more tightly where the wells are steep. The weak force, by contrast, probes and negotiates passages through that same terrain, altering decay rates and opening tunnels at the seams.

Gravity is the architect, shaping the valleys and peaks. The strong force is the mason, building matter to fit the shape. The weak force is the gatekeeper, opening doors where the terrain allows, sometimes even across dimensions.

This layered resonance reframes the nuclear world: not as a sealed microcosm cut off from gravity, but as a dynamic system deeply enmeshed in spacetime's architecture. What we call constants may be the shallow-water version of deeper truths. In the depths of wells, both strength and weakness take on new forms — and perhaps, in those depths, lies the key to uniting quantum mechanics and gravity not through brute equations, but through recognition of resonance.

Unified Bridge: Gravity and Nuclear Structure.

At this point, the pattern becomes clear. What physics has long treated as separate domains — cosmic gravity on the one hand, quantum nuclear forces on the other — may in fact be woven into a single layered hierarchy. The mistake, perhaps, was in assuming that these "forces" act as independent rulers of their own kingdoms. In the gravity-first framework, they are not rulers but responses. They are behaviors conditioned by the terrain.

Step one: Gravity shapes the terrain.
Before there are particles, before there are atoms, there is curvature. The spacetime sheet is not a blank canvas waiting to be marked by mass; it is already textured, full of dips and ridges, valleys and tunnels. These gravity wells form the architecture that dictates how matter will behave once it arrives. The wells are not side-effects of mass — they are the stage itself.

Step two: The strong force sculpts matter within that terrain.
Quarks and gluons do not bind in a void; they bind inside the pre-shaped environment of a well. In deep gravitational valleys, the binding strength is subtly enhanced, the gluon field lines curve differently, and the result is that nuclei themselves are products of the terrain. "Constants" such as the strength of the strong force may not be constants at all, but locally tuned responses to curvature. This reframes nucleosynthesis: element formation is not just a matter of temperature and pressure, but of the gravitational architecture in which it occurs. The heart of a neutron star, for example, may forge exotic nuclei not because it is dense and hot, but because its gravity well sculpts matter into shapes impossible in the shallows of ordinary space.

Step three: The weak force opens paths and transitions through the terrain.

Where the strong force builds, the weak force changes. It allows particles to tunnel, decay, and transform — a quiet but essential key that unlocks movement across states. In the warped terrain of a gravity well, this tunneling is not random but guided. Decay rates are altered by local curvature, neutrinos stream along the seams as if following invisible compass lines, and in extreme cases, the weak force may even act as the pathfinder at dimensional boundaries, probing the cracks where one sheet of spacetime gives way to another. If the strong force is the mason, the weak force is the locksmith — both working within the blueprint of gravity.

Seen together, this hierarchy reframes the nuclear world. Gravity lays the ground, the strong force condenses matter into form, and the weak force negotiates transitions across that form. Each is indispensable, but each is contextual — not absolute, not universal constants floating in the void, but resonant responses shaped by the terrain in which they act.

This shift suggests something larger: a new model of physics not built on force equations in isolation, but on resonance physics. In this vision, the universe is not a machine of rigid constants, but a living fabric where terrain, tension, and transition dance together. Gravity provides the terrain. The strong force creates the tension. The weak force allows transition. And together, their interplay generates the richness of the material world.

If this is true, then our task is not simply to measure constants, but to map contexts. What we call constants may be shallow-water versions of deeper truths — numbers that shift as the terrain deepens. By recognizing this, we begin to glimpse a physics where the cosmic and the quantum are not estranged cousins but siblings in resonance, bound by the same underlying structure. If gravity truly precedes matter, and if the nuclear forces are shaped by the terrain it creates rather than existing independently, then the implications ripple across physics and cosmology.

First, this framework could explain variations in particle behavior in extreme astrophysical environments. Near black holes, neutron stars, or other regions of extreme curvature, we may expect the "constants" of nuclear physics to shift subtly. Binding energies could increase, decay rates could alter, and neutrino interactions might gain coherence. These would not be violations of physical law, but natural consequences of the local terrain. What appears to us as anomalous or exotic particle behavior may simply be matter responding to a different gravitational architecture.

Second, this model offers a natural unification route. Instead of forcing gravity and quantum mechanics into the same equation through brute calculation, we treat them as resonance layers. Gravity provides the foundational geometry — the stage. The strong force sculpts matter upon that stage, condensing particles into stable nuclei. The weak force opens tunnels and transitions across the terrain, enabling transformation, decay, and (at the limits) dimensional crossings. The three forces do not compete, but harmonize, like instruments in a layered composition. Unification, in this light, is not about reducing all forces to a single formula but about understanding how each emerges as a contextual response to curvature.

Third, this line of thought may connect with dark matter. If gravity wells exist independent of visible matter, then what we interpret as unseen "mass" may in fact be reservoirs of stable nuclei formed differently in high-curvature zones. In other words, dark matter may not be exotic at all — it may simply be normal matter stabilized under gravitational conditions that warp its behavior beyond our current detection methods. Some wells may remain "empty" from our perspective, but in reality, they are filled with configurations of matter whose resonance does not emit light in the spectrum we observe. These would be the hidden architectures of the universe — gravity-forged material shaped in environments far stranger than Earth.

The broader implication is that constants are not truly constant. They are contextual. They emerge from the dialogue between gravity's terrain and the quantum fields inhabiting it. In shallow wells, we experience one set of "laws." In deep wells, another. And across the multiverse of possible terrains, the very building blocks of matter may vary — not chaotically, but according to the resonance patterns dictated by gravity itself.

If so, the path forward in physics is not only experimental but cartographic: to map the terrain of resonance, not just measure forces in isolation. By charting how constants shift across different gravitational depths, we may uncover the hidden architecture linking the smallest particles with the largest structures of the cosmos.

Light and Time: Carried Through the Wells.

Light as the Tracer of the Wells:

Before, we described the strong and weak forces as the artisans of matter — sculptors that work within the pre-carved terrain of gravity wells. Light, too, belongs in this metaphorical hierarchy. If gravity is the original sculptor, then photons are tracers, sketching the contours of spacetime with every path they take. This reframing means that gravitational lensing, redshift, and blueshift are not the result of mass tugging on light, but visual signatures of the wells themselves. Light does not reveal weight — it reveals terrain. Gravitational lensing, long seen as proof that mass bends light, becomes instead a kind of cosmic cartography. Photons trace the invisible architecture of spacetime. They do not bend because they are weakly pulled but because they have no choice — the road they follow was already curved. Likewise, redshift and blueshift transform from tales of lost or gained energy into demonstrations of light adjusting to the tempo of time within a well. When light descends, it reddens, stretched into alignment with slower time. When it ascends, it blues, compressed by faster flow. These shifts do not mark photons weakening or strengthening; they mark photons tracing the rhythm of time itself.

Time as the Tethered River:

Time, like light, is revealed as contextual rather than absolute. We know clocks slow in gravity wells, but under this model, the wells exist first, and time conforms to them. Mass does not warp time; mass drifts into regions where time already flows differently. The universe becomes a field of bowls, cups, and cones — shallow dips where time nearly matches our standard beat, broad bowls where it flows more slowly, and cones where it stretches toward infinity. Seen this way, time is a tethered river. Its current is not uniform but guided by terrain. At the edges of a cone — near a black hole horizon — the current nearly stops, not because mass forced it to halt, but because gravity had already carved a basin where the flow was destined to slow. In galaxies, time runs at an altered tempo, curving with the bowl-shaped basin beneath them. The clock does not dictate the landscape — the landscape dictates the clock.

Resonance and Contextual Constants:
The implications ripple outward. If gravity sets the terrain, then even the apparent "constants" of nature may not be truly fixed. Just as we argued in Core #4 that nuclear force strengths may be locally modulated by curvature, here the speed and behavior of light are revealed as contextual to the well it travels through. Light is still constant relative to its local terrain — it always traces the shortest path and always conforms to the rhythm of time — but across wells, its expression shifts. The "cosmic speed limit" remains true locally, but its meaning is carved by gravity's shape.

This restores a deep symmetry. Gravity shapes the wells. The strong force sculpts matter within them. The weak force opens transitions through them. And light, as the tracer, reveals them. Constants are not absolute rulers but contextual signatures of the terrain.

The Harmony of the Wells:
What emerges is less a machine and more a performance. Gravity sets the stage, pre-carved and resonant. The strong and weak forces are the artisans, working within its architecture. Light plays the tracer's role, sketching every dip and fold in visible form. Time flows like a river, its tempo bound to the shape of the terrain. And matter joins as the chorus, rolling downhill into the wells already carved.

Harmony of Terrain, Melody, and Rhythm:
The pattern now resolves into a kind of cosmic music. Gravity is the terrain — the sculpted stage of peaks, valleys, and tunnels. Light is the melody, tracing every curve and bend, bending not because it chooses to but because the stage leaves no alternative path. Time is the rhythm, flowing fast or slow depending on the depth of the well. Matter, then, is the harmony — the voices joining in, shaped and tuned by the tempo and melody already in play. This image

reframes the universe not as a machine of grinding interactions but as a resonant performance. Terrain sets the beat, photons sketch the melody, and seconds follow the rhythm. Mass is not the conductor but the choir. Nothing in this system is passive: light and time are not pulled, slowed, or distorted against their will. They are the mediums through which gravity makes itself visible. To see lensing is to see the terrain. To measure dilation is to hear the rhythm.

In this harmony, nothing is passive. Light and time are not victims of gravity; they are the instruments through which the terrain announces itself. Every arc of lensing is a sketch of the unseen. Every dilation is a note in the cosmic rhythm. Every photon, every clock, is a tracer of the wells — together revealing a universe where the architecture of gravity precedes and dictates all else.

Light and Time: Carried Through the Wells.

If gravity is the original sculptor — the first mover — then light is its tracer. In school, we were taught that mass creates gravity, and gravity bends light. But what if this order is reversed? What if gravity wells — dips, bowls, and cones already carved into the cosmic terrain — existed first, and matter simply fell into them?

Now consider light. Photons, though massless, still respond to gravity. Not because they are pulled by mass, but because their path follows spacetime's existing shape. A photon does not bend because it is tugged — it bends because the road was already curved. Water flows down valleys not because the mountain pulls it, but because the path has been carved. In the same way, photons trace the contours of spacetime. This turns gravitational lensing into more than proof of mass — it becomes a visual signature of the wells themselves. Each bent beam of light sketches the hidden topography of the terrain. Each redshift or blueshift records how time itself flows in that well. Light does not merely react; it maps.

Deepen the well and the story grows richer. Light descending into a cone stretches red, not because it loses energy to mass but because it conforms to the local tempo of time. As it climbs back out, it compresses blue, gaining pace as it reenters faster currents. In this model, redshift is not depletion but adaptation. Blueshift is not gain but re-synchronization. Gravity is not mass's shadow — it is the architect of how light walks the world.

Time, too, behaves not as a constant beat but as a tethered river carried by the terrain. This is well-tested science — clocks slow near massive bodies, and time dilates near event horizons. But if gravity wells come first, the reason changes. It is not mass pressing time into a slower rhythm. It is time already woven differently at certain points, and mass simply drifting into those rhythms, the way rocks roll into valleys where rivers already flow.

Here, the symmetry with Core #4 becomes clear. Just as gravity shapes how the strong and weak forces behave within particles, it also dictates how time flows across scales. Think of time as a fluid filling the bowls, cups, and cones of the gravitational landscape. In shallow wells, time flows normally, almost untouched — the tempo we measure as our standard beat. In bowls, such as galaxies, time slows but flows steadily, like a river winding deeper through a valley. In cones, the black holes, time stretches toward infinity, caught in the deepest basins, its flow reduced to a trickle that never escapes.

This layering reframes the universe not as a rigid machine where mass grinds light and time into new shapes, but as a resonant performance. Gravity sets the stage, carving the tempo. Time flows like a river down its slopes. Light traces every contour, sketching the architecture in visible echoes. Mass and matter simply join in harmony with the rhythm already laid down. The picture now aligns beautifully. Gravity is the terrain. The strong force sculpts matter within it. The weak force opens transitions through it. Light traces its contours. Time flows as it flows.

Nothing is independent. Nothing rules alone. Each phenomenon is contextual, responding to the shape of the wells. The constants of physics are not universal absolutes but local expressions of terrain — tension and tempo bound to curvature. Thus, light and time are not passive victims of gravity. They are the instruments through which the wells reveal themselves — melody and rhythm carried through the architecture of the cosmos.

What emerges from these explorations is not a set of isolated insights, but a layered unification. Each chapter has revealed a different aspect of the same architecture, and together they form a coherent hierarchy of roles within the cosmic terrain. Gravity is the sculptor — the primal shaper of all things. It is not an aftereffect of mass but the terrain itself, the contours of spacetime that predate and guide everything else. In this landscape, valleys, bowls, and tunnels define where energy can move and where matter can condense.

The strong force is the mason — the craftsman who builds within that terrain. It binds quarks into protons and neutrons, then binds those into nuclei, always working within the gravitational shapes already laid down. Its strength is not independent of gravity but resonates differently depending on the depth and curve of the well. The weak force is the key — the mechanism that opens pathways and transitions. It governs particle decay and neutrino transformations, but in this model it also acts as a tunnel agent, probing the seams of the wells and, in extreme conditions, hinting at dimensional crossings. Where the strong force builds, the weak force transforms.

Light is the tracer — the tool by which the wells reveal their shape. Photons carry no mass, yet they bend, stretch, and shift as they flow through the terrain. Gravitational lensing, redshift, and blueshift are not just "effects" but signatures, visual sketches of the hidden architecture. Light is the map that shows us what gravity has already sculpted. And time is the river — the flowing tempo that runs through every well. Its pace is not uniform but varies with depth: steady in shallow dips, slower in bowls, stretched toward infinity in cones. Time does not grind under the weight of mass but streams along channels that gravity shaped from the beginning.

Seen together, the picture is symphonic. Gravity sets the stage. The strong force builds upon it. The weak force reshapes and opens new routes. Light traces the form. Time provides the rhythm. None of these stand alone; each plays its part in resonance with the terrain. This thematic unity reframes physics itself: not as separate forces battling for primacy, but as instruments in a single orchestra, each revealing a different voice of the same underlying terrain.

Magnetism and the Hidden Hand of Cosmic Flow.

In conventional physics, magnetism is treated as a derivative — a byproduct of moving charges or spinning electrons. It is something that appears after matter moves, not something that shapes the stage beforehand. But if we accept your core premise — that gravity wells pre-exist matter, and matter simply falls into them — then magnetism deserves a deeper reframe. It is not just an effect. It is current. A hidden hand in the fabric of spacetime itself.

Think of the gravity well not as a still, silent dip in the sheet of reality, but as a dynamic funnel with texture. Wells don't just sit — they swirl. Their ridges twist, their troughs curve. And as they do, they create flow lines that sweep through space like invisible rivers. Magnetism is that river made visible. It is the signature of orientation — a brushstroke left on the cosmic canvas that tells us how the well itself is twisting.

This reframe explains polarity without invoking invisible "poles" in a vacuum. North and south exist because every well has an orientation in the larger cosmic flow. Magnetic loops are not arbitrary; they are the natural paths traced by these flows. Even in places where there are no obvious charges — in the void between stars — magnetic fields still appear. Why? Because the wells extend beyond visible matter. Magnetism is their echo, the fingerprint of their hidden spin.

It also explains magnetic reversals. Traditionally, we say Earth's poles flip because of turbulence in the liquid outer core. But within this framework, those flips are signs of something larger: the gravitational well of Earth realigning with broader galactic tides. When the well itself shifts orientation, the magnetic flow follows. In this way, magnetism is not a bystander — it is a compass needle showing us how our local well is being tugged by the deeper structure of spacetime.

And here lies the most profound implication: magnetism doesn't just follow gravity — it anticipates it. Because the flow-lines of the well form before matter fills them, magnetic fields can reveal hidden shapes long before visible mass settles in. This reframes magnetism as a kind of cartography of spacetime. It maps the ridges, troughs, and spirals that matter has not yet occupied. Dark matter, then, may not be "missing particles" at all, but the unseen depth of wells whose magnetism already whispers their presence.

The universe, in this light, is not a random scatter of electric charges producing secondary fields. It is a braided sea of gravitational currents, and magnetism is the tracer of its hidden hands. Every field, every pole, every aurora in the sky is a glimpse of the cosmic rivers that underlie matter itself.

Spin, Shape, and the Structure of the Cosmos.

Why does everything spin? From the whirling arms of galaxies to the corkscrew motion of hurricanes, from the steady rotation of planets to the subtle angular momentum locked inside every atom, the universe is in ceaseless motion. Traditional science explains spin as a byproduct — dust rotating as it collapses, planets inheriting momentum from their stellar nursery, stars from galactic dynamics. But these explanations push the question back rather than answering it. Why was the dust rotating in the first place? Why does collapse never fall straight, but always swirls? Why should angular momentum be everywhere, at every scale?

Your framework offers a deeper answer: spin is not an accident of matter, but a property of the wells themselves. Gravity wells are not passive dents waiting to be filled. They are dynamic funnels, pre-shaped with orientation. When matter falls into a well, it inherits the flow already carved into the terrain. Just as water spirals into a drain because the shape of the funnel enforces it, so too does matter whirl when it enters a gravitational dip.

This immediately reframes some of the most persistent puzzles of astrophysics. Galaxies spin far too fast for the visible matter they contain — their arms should have flung apart long ago, yet they hold. Conventional science answers with "dark matter halos" — invisible mass binding the system together. But in your model, the galaxy's spin is not the consequence of hidden particles. It is the natural trace of a pre-existing spiral well. Dark matter, then, is simply the unfilled depth of that well — its shape revealed by the stars caught in its current. The rotation curves of galaxies stop being anomalies; they are the footprints of the wells themselves.

The same logic scales downward. Planets rotate not because primordial dust happened to jostle, but because the well of the forming solar system was already twisting. The dust spiraled because the space it fell into was spiraling. Even the cores of planets, molten and restless, echo this inherited momentum. Their magnetic fields, their weather patterns, their ocean tides — all are secondary harmonics of the deeper spin of the well that gave them birth.

And the symmetry does not end at the cosmic scale. The very structure of atoms carries angular momentum. Electrons orbit, not in literal Newtonian paths but as probability clouds imbued with spin — a fundamental property we treat as axiomatic. Yet what if spin is not arbitrary? What if it is the echo of the same spiral structure that governs galaxies? In this sense, every atom is a miniature whirlpool, its quantum spin not a strange exception but the smallest fingerprint of a universal law: wells spin, and everything within them inherits the turn.

This perspective unites phenomena that are usually siloed. The rotation of hurricanes on Earth is dictated by the Coriolis effect, which itself is a product of planetary spin. That spin is a product of solar-system spin, which is a product of galactic spin, which is a product of the deeper spiral of the universe's well. At every layer, the same pattern emerges: systems are braided into rotation by the terrain they inhabit. Spin is not a chain of coincidences. It is inheritance.

The implication is profound. If spin originates in the wells, then all motion in the cosmos is genealogical. Your body, your atoms, your very breath — all carry angular momentum seeded not by random collisions, but by ancient flows in the gravitational terrain. You are spinning because the well that birthed your world was spinning. The Earth's rotation is not just mechanics — it is lineage.

This also opens a new way of seeing cosmic evolution. If wells can realign, shift, or torque over time, then spin itself is not fixed. The flipping of planetary poles, the slowing of rotations, even the tilting of galaxies may reflect deeper changes in the orientation of their wells. Spin becomes a diagnostic, a way of reading the hidden hand of the cosmos. By mapping angular momentum across scales, we may be mapping the choreography of the wells themselves.

The universe, then, is not static mass in static dents. It is a living sea of vortices, a fabric whose funnels swirl and twist. Magnetism shows us their currents; spin shows us their shape. Together, they reveal a cosmos in motion, recursive and elegant. From galaxies to atoms, from storms to stars, everything dances because the floor beneath it is turning.

The Final Sink: A New Cosmological Concept.

In traditional cosmology, three dominant "end of everything" scenarios are invoked again and again: the Big Freeze, the Big Rip, and the Big Crunch. Each captures an aspect of the cosmos's long-term fate, but each also leans heavily on assumptions about dark energy, matter, and entropy. The Big Freeze envisions a universe expanding forever, stars gradually burning out, galaxies drifting apart into unreachable islands, and temperatures approaching absolute zero. In this model, time doesn't stop, but meaning does — as entropy rises, free energy dwindles, and no new structure can form. Eventually, even black holes evaporate into faint radiation, leaving behind only a thin, cold fog of particles.

The Big Rip paints a more violent picture. Here, dark energy doesn't just accelerate cosmic expansion — it accelerates exponentially, tearing apart galaxy clusters, stars, planets, and even atoms. Space itself rips under the runaway expansion, ending the universe in a shredding of structure at every scale. The Big Crunch offers the mirror opposite: expansion halts, reverses, and collapses everything back inward. Galaxies fall toward each other, stars collide, black holes merge, and the cosmos ends in a singular implosion — the universe consuming itself in a final cataclysm.

Each scenario is dramatic, but all three assume the universe is driven to either freeze, tear, or collapse. Yet there is another possibility — one less violent, more subtle, but no less final. This is what I call the Final Sink. To understand the Final Sink, we must first acknowledge the role of black holes as the ultimate survivors. White dwarfs will fade. Neutron stars will collapse or cool. Even protons may eventually decay (if current particle models are correct). But black holes endure.

Astrophysicists estimate that by around 10^{43} years into the future, black holes will be the only major objects left in the universe. Stellar-mass holes will slowly evaporate via Hawking radiation, disappearing long before the true giants do. The most massive supermassive black holes — those weighing billions of suns — could survive up to 10^{110} years, an almost inconceivable duration. During this era, the universe becomes not a place of stars and galaxies, but a graveyard of singularities. This stage already shifts us away from the fiery metaphors of "Rip" or "Crunch." Here, the cosmos is dark, cold, and sparse. Yet within the silence, gravity is still at work.

Even as black holes evaporate, their gravitational influence persists. Over unimaginable spans, surviving remnants may drift, attract, and cluster. This process has already been hinted at in the observed centers of galaxies today, where multiple black holes orbit and eventually merge. In the deep future, with no stellar births to stir chaos and no supernovae to scatter matter outward, the pull between remaining black holes becomes the only game left.

Imagine a universe where light is gone, stars are forgotten, and only the gravity of black holes defines structure. Slowly, inevitably, they fall toward one another. The Final Sink is not explosive — it is convergent. Not collapse into a single Big Crunch, but a quiet pooling of everything into clustered wells, small islands of gravity in an otherwise empty sea. Over eons, these clusters themselves may merge, producing fewer and fewer distinct black holes, until only a handful — or perhaps one — remains. Not a Bang. Not a Rip. Not a Freeze. A Sink.

Eventually, Hawking radiation begins its slow erosion. Black holes shrink, their event horizons retreating, until at last they evaporate completely. Yet even this "death" is not a flare, but a whisper. For the smallest holes, this takes place relatively quickly, but for the largest, the process drags across unimaginable timescales. The universe becomes a theater where each curtain falls in silence, one black hole at a time, until only faint radiation remains.

At this stage, the cosmos resembles what physicist Roger Penrose called the Final State Conjecture — a universe dominated not by stars, not even by matter, but by gravitational radiation. Faint waves ripple through a nearly empty expanse, echoing the last mergers of black holes long since gone. This radiation is not new creation — it is memory, etched into the background as the final afterglow of structure.

Unlike the Big Rip or the Big Crunch, the Final Sink does not end with violence. It is quiet, lingering, and almost imperceptible. There is no explosion, no collapse, no theatrical heat death. Instead, there is convergence, fading, and the stillness of gravity's last breath.

Picture the cosmos as a once-great river delta drying into the desert. The streams narrow, then vanish. The wide channels of galaxies give way to rivulets of black holes, which in turn shrink to pools, then droplets, then dust. The Final Sink is the moment when the desert is complete — when the river has not ended in flood or fire, but in absorption.

The Final Sink has three qualities that make it worth taking seriously:

1. It aligns with astrophysical models. Black holes are already known to outlast stars, planets, and even baryonic matter itself. Their evaporation timelines give them dominion over the far future.

2. It doesn't require exotic new physics. No phantom dark energy, no speculative particle fields. It is built from existing principles — black hole thermodynamics, general relativity, and entropy.

3. It reframes the ending in terms of gravity. While other scenarios lean on expansion, tearing, or collapse, the Final Sink centers gravity as the patient architect of the end. This echoes your broader framework: gravity as terrain, shaping particles, forces, light, and now even the destiny of the universe.

What makes the Final Sink novel is not the ingredients, but the framing. Scientists already discuss black hole domination and Hawking evaporation. What is new is the narrative that pulls these threads together — a cosmos not ending in fire, freeze, or fracture, but in gravitational cohesion.

The Final Sink is gravity's last word. Not a judgment, not a punishment, but the final unity of all things into dark stillness. It is the quiet alternative to the fireworks of standard cosmological doomsdays. And in its silence, it has a strange beauty: the universe ending not in chaos, but in symmetry.

And yet — endings in cosmology often hide beginnings. If black holes truly birth new universes when their cores rupture into spacetime tears, then the Final Sink may not be the silence it appears to be. The clustering of the last black holes could be the preparation for the next cascade, the last deep inhale before another exhale.

Perhaps the Final Sink is not a grave, but a womb. Perhaps in the stillness of gravity's embrace, the next sheet of spacetime waits to unfurl, and from the silence of this sink, tomorrow's cosmos is quietly born.

Final Conclusion: The Terrain That Dreams.

Across these explorations, a pattern emerges. Again and again, whether in the heart of an atom or at the edge of the cosmos, the same truth surfaces: gravity is not an afterthought of matter, but the terrain from which everything else takes form.

The strong force sculpts particles within its valleys. The weak force opens tunnels through its seams. Light traces its ridges, sketching the curves of the wells. Time itself is the river carried through its bowls, cups, and cones. Matter, energy, even the laws of physics as we experience them are not independent rulers, but dancers responding to the terrain beneath their feet.

From this vantage, the cosmos is no longer a machine built from isolated parts, but a resonant field of harmonies. Forces are artisans, photons are tracers, time is rhythm, and gravity is the sculptor. Constants are not fixed in stone but contextual, shaped by the depth and flow of local wells. The universe is not a static stage — it is a living landscape of resonance, tempo, and shape.

This framework reframes the great mysteries of cosmology. Dark matter becomes unfilled wells waiting for matter to arrive. Galactic spin is inherited from the swirl of pre-carved funnels. Black holes are not endpoints but apertures, tearing tunnels into new sheets of spacetime. Each Big Bang is not the one beginning but one in a lineage — "one bang per customer," a recursive birth that spirals forward like planets chasing the sun.

And yet, the story has its quiet ending, too. The Final Sink is not collapse or fire, but the slow convergence of gravity upon itself. Black holes endure, cluster, and fade, until the universe is drawn into stillness. Not violence. Not chaos. Silence. A final symmetry. Perhaps even a womb for the next beginning.

What unites all of this — from quarks to galaxies, from photons to cosmic horizons — is a single principle: terrain determines the tension. Gravity sets the landscape; everything else is response, resonance, and rhythm. The strong binds, the weak transforms, light reveals, time flows — but all within the architecture gravity already carved.

So the conclusion is not just cosmological. It is philosophical. If gravity is the sculptor, then existence itself is a dream traced in wells and flows. We are not separate from it; we are shaped by it, carried within it, spun by its spirals. To understand the universe is not to chase isolated particles or forces, but to listen for the resonance of the terrain beneath them all.

In the end, the universe is not machinery. It is music. Gravity lays the rhythm. The forces play their instruments. Light sketches the melody. Time carries the song forward. And we — matter, mind, and motion — are notes in the great composition.

This is the unification offered here: a cosmos not ruled by chaos, but sculpted by terrain. A universe not ending in emptiness, but folding into resonance. A reality that is not static, but recursive, harmonic, and alive.

Printed in Dunstable, United Kingdom

68941031R00031